Leveraging Complexity in Great-Power Competition and Warfare

Volume I, An Initial Exploration of How Complex
Adaptive Systems Thinking Can Frame
Opportunities and Challenges

SHERRILL LINGEL, MATTHEW SARGENT, TIMOTHY R. GULDEN,
TIM MCDONALD, PAROUSIA ROCKSTROH

Prepared for the Department of the Air Force
Approved for public release; distribution unlimited

PROJECT AIR FORCE

For more information on this publication, visit **www.rand.org/t/RRA589-1**.

About RAND

The RAND Corporation is a research organization that develops solutions to public policy challenges to help make communities throughout the world safer and more secure, healthier and more prosperous. RAND is nonprofit, nonpartisan, and committed to the public interest. To learn more about RAND, visit www.rand.org.

Research Integrity

Our mission to help improve policy and decisionmaking through research and analysis is enabled through our core values of quality and objectivity and our unwavering commitment to the highest level of integrity and ethical behavior. To help ensure our research and analysis are rigorous, objective, and nonpartisan, we subject our research publications to a robust and exacting quality-assurance process; avoid both the appearance and reality of financial and other conflicts of interest through staff training, project screening, and a policy of mandatory disclosure; and pursue transparency in our research engagements through our commitment to the open publication of our research findings and recommendations, disclosure of the source of funding of published research, and policies to ensure intellectual independence. For more information, visit www.rand.org/about/principles.

RAND's publications do not necessarily reflect the opinions of its research clients and sponsors.

Library of Congress Cataloging-in-Publication Data is available for this publication.

ISBN: 978-1-9774-0758-0

Cover: Olivier Le Moal/Getty Images/iStockphoto.

Limited Print and Electronic Distribution Rights

Preface

Modern warfare is characterized by a complex environment of operations spanning multiple domains leveraged to achieve advantage against the adversary in both strategic competition and armed conflict. However, understanding complexity in warfare, often referred to as the *art of war*, is rarely framed in a structured way. This report seeks to provide an initial examination of how complex adaptive systems thinking can frame opportunities and challenges of complexity in warfare. This report is the first in a two-volume report. The second volume, *Leveraging Complexity in Great-Power Competition and Warfare*: Volume II, *Technical Details for a Complex Adaptive Systems Lens*, provides additional information to support this volume. This report examines how complex adaptive systems thinking can be applied to great-power competition and warfare to aid in understanding how complexity might be exploited to U.S. advantage. This report should be of interest to planners, geographic commands, and the military science and technology community.

The research reported here was commissioned by the Air Force Research Laboratory executive director, Jack Blackhurst, and conducted within the Force Modernization and Employment Program of RAND Project AIR FORCE as part of a fiscal year 2019 add-on project Complexity Imposition.

RAND Project AIR FORCE

RAND Project AIR FORCE (PAF), a division of the RAND Corporation, is the Department of the Air Force's (DAF's) federally funded research and development center for studies and analyses, supporting both the United States Air Force and the United States Space Force. PAF provides DAF with independent analyses of policy alternatives affecting the development, employment, combat readiness, and support of current and future air, space, and cyber forces. Research is conducted in four programs: Strategy and Doctrine; Force Modernization and Employment; Manpower, Personnel, and Training; and Resource Management. The research reported here was prepared under contract FA7014-16-D-1000.

Additional information about PAF is available on our website: www.rand.org/paf/

This report documents work originally shared with DAF in March 2020. The draft report, also issued in March 2020, was reviewed by formal peer reviewers and DAF subject-matter experts.

Contents

Figures

Summary

Issues

A key concern for the U.S. Air Force is the ability to leverage multidomain operations (MDOs) to U.S. advantage both in competition and in warfighting. Multidomain actions are viewed as imposing complexity on the adversary's decision process. But what does *complexity* mean in an Air Force context?

- Currently, we lack understanding of how to impose or exploit complexity to maximize operational effects.
- An informed understanding of the nature and value of complexity as a weapon is necessary—that is, how complexity can be used as a method of attack. How can complexity be employed in an operational setting?
- Science and technology investments are not currently aligned to quantify complexity, measure its operational effects, and determine how to impose complexity and thus shape adversary behavior. How can research on the nature of complexity be used to understand what science and technology efforts might deliver in terms of complexity-informed capabilities?

Approach

The RAND Corporation research documented in this report involved a literature review of complexity, graph theory, and complex adaptive system (CAS) to ground the complexity characterization in research relevant to major interstate competition and warfare. The warfighting application of the RAND-developed complexity lens can be applied to four example vignettes, which leverage the emerging MDO concept of operations (CONOP) from recent wargames and a review of historical case studies. The focus on *complexity imposition* or *complexity exploitation* in our analysis is on an adversary's decisionmaking because the approach influences the decision calculus to arrive at strategic or operational decisions.

Conclusions

- To impose or exploit complexity is to take an action that increases an aspect of the complexity of the environment in a way that makes it more difficult for an adversary to make decisions or to operate, essentially shaping conditions to favor Blue. Thus, to conduct a complexity attack is to take an action that exploits characteristics of CAS in a way that will have a deliberate negative effect on the adversary.
- Applying a CAS lens to warfare is useful to understand how a planner might leverage complexity to U.S. advantage.

- Four general categories of U.S. actions benefit from the CAS nature of the adversary's decision processes. These are degrading the operational picture, impairing adversary response, spanning organizational boundaries, and exploiting nonlinearities.
- The difference between complexity exploitation and more-traditional actions are that the complexity-based response reshapes an adversary's decision calculus, while more-traditional options simply lower the probability associated with a given transition from one decision step to the next.

  - There may be cases in which a given course of action exhibits both types of impact.
  - The value of this method is that it provides a useful definitional frame that highlights the unique aspects of complexity-based attacks, and it allows the impacts of different types of Blue actions to be directly compared with one another in terms of their ultimate effects on an adversary's decisionmaking.

- U.S. actions on the adversary's decision points in a military vignette can be represented by CAS characteristics that are represented in a Markov chain.

  - Understanding the underlying transition matrix structure (or associated directed graph) tells us some fundamental things about complexity (e.g., the creation of feedback loops).
  - Developing a model with defined probabilities, while challenging, may provide additional insights for warfighting.

Recommendations

- The Air Force should apply a complexity lens through which to review ongoing and future efforts to best leverage complexity to U.S. decision advantage. The lens should take both offensive (opportunities for attack) and defensive (highlight and address vulnerabilities) looks at systems. The efforts under review should span

  - science and technology needed to leverage complexity
  - planning for MDOs
  - evaluation of MDO effectiveness.

- The Air Force Research Laboratory should conduct game theory–informed wargames to populate probabilities in the mathematical representation of the adversary's complex decision process.
- The Air Force should conduct workshops in support of Joint All-Domain Operations concepts, leveraging the complexity lens to inform CONOP development.
- Pacific Air Forces and U.S. Air Forces in Europe should integrate complexity lens thinking into their existing tabletop and command post exercises.

Acknowledgments

We thank the Air Force Research Laboratory Information Directorate sponsors, Lee Seversky and David Myers, thought leaders on command and control and complexity who were extremely engaged with the RAND team on this project. This work would not exist without their intellectual curiosity and patience. We benefited greatly from the participation of our colleagues, who provided initial feedback on CAS and multidomain warfare in two workshops: Jim Chow, Paul Davis, Dave Frelinger, Tom Hamilton, Karl Mueller, Angela O'Mahony, Dave Ochmanek, Dave Shlapak, Michael Spirtas, and Rand Waltzman. We also thank Caitlin McCullouch, our RAND summer associate, for her historical case study work, which helped clarify our thinking on complexity in warfare.

Abbreviations

ACC	Air Combat Command
ACE	agile combat employment
AFRL	Air Force Research Laboratory
C2	command and control
CAS	complex adaptive system
CONOP	concept of operations
IADS	integrated air defense system
ISR	intelligence, surveillance, and reconnaissance
MDO	multidomain operation
OODA	observe, orient, decide, act
SAM	surface-to-air missile
SEAD	suppression of enemy air defenses
S&T	science and technology
TRADOC	Training and Doctrine Command

1. Complexity Imposition: The Hypothesis

Introduction

The Air Force 2035 Future Operating Concept calls for planning and executing multidomain operations (MDOs) in an evolving battlespace with speed and scale through mastery of command-and-control (C2) complexity (Air Force, 2015). Mastering complexity of U.S. C2 is needed to effectively compose, orchestrate, and apply force across all domains. The United States must transition from the *current state* of loosely coupled, handcrafted multidomain plans and assessment to a *future state* of intertwined, continuous planning, execution, and assessment. There is also another side of complexity. The adversary is similarly faced with an evolving battlespace, with decisionmaking occurring in an increasingly complex C2 system. The complexity that the adversary faces in conflict and competition with the United States could be seen as opportunities for U.S. advantage. Although a critical part of operational complexity, how we use and manage complexity of U.S. C2 is not the focus of this effort. Instead, we focus on using complexity to make it harder for the adversary to exercise effective C2 of its forces, thus creating U.S. advantage.

Mastering C2 complexity requires understanding and quantifying the desired outcomes of imposing complexity on the adversary. Historical accounts of the effects of deception and confusion on the battlefield (i.e., the fog of war) at the tactical, operational, and strategic levels abound and represent the art of warfare. A complexity attack on the adversary may exacerbate the fog of war by creating conditions of insufficient knowledge, uncertainty of one's own strength and positions (and that of the opponent), differences between expectations and reality, logistic difficulties, and the need to replan when confronted with the battlefield environment.[1] Aside from exacerbating the fog of war, a complexity attack can simultaneously present the adversary with a multitude of dilemmas that obviate any simple or singular response (i.e., the horns of multiple dilemmas).[2] The effects of a complexity attack arise, then, from placing the adversary in challenging impasses while denying it information needed to resolve the impasses. Although the United States can impose complexity on its adversaries, they can do the same to the

---

[1] Complexity as a weapon should be considered from both offensive and defensive points of view. Although offense and defense require the same fundamental understanding, offense and defense each demand their own sets of tools. The focus of this effort is the offensive application.

[2] *Dilemmas* refer to contexts that require a choice between two or more undesirable options. By simultaneously presenting multiple threats and a perceived absence of concrete information, such scenarios are particularly difficult for commanders to navigate because their decisionmaking is strongly conditioned by cognitive biases, such as loss aversion and the desire to mitigate immediate sources of risk even if they are strategically counterproductive (e.g., zero-risk bias). The lack of an optimal course of action and a psychologically conditioned focus on the immediate costs will often slow decisionmakers and cause them to optimize on immediate risk reduction rather than maximizing long-term advantage (Rosenthal, Boin, and Comfort, 2001).

United States. We should, therefore, be equally concerned with ways in which the United States can defend against adversary complexity attacks, although this is not the focus of this report.[3]

The purpose of the RAND Corporation's two-volume series is to provide some structure to the art of warfare in terms of complexity and thereby help warfighters think about how to use complexity as a weapon to gain operational advantage over the adversary. The research provides examples of leveraging complexity characteristics against an adversary's decisionmaking process. It is an initial exploration of complex adaptive system (CAS) thinking in warfare and not a definitive document.

Why might it be important to think about modern warfare in terms of complex systems? In the extreme, amassing forces in sufficient numbers regardless of how *complex* the adversary and environment are could obviate the need for this discussion. However, given the fact that military force structures are constrained by budgets and other resource concerns, it is not a given that this option is possible. Furthermore, today's modern warfare places a premium on what China terms *informationalized warfare* and is pervasive in People's Liberation Army writing about systems confrontation and systems destruction warfare. Therefore, we find that viewing the adversary's decisionmaking process as a CAS to be a promising way to think about warfare (Engstrom, 2018).

Currently, we lack understanding of how to exploit complexity to maximize operational effects. More generally, we lack a framework for quantifying and scientifically studying the effects of complexity on cognitive load and decisionmaking—precursors to applying complexity in an informed and deliberate manner. A logical start may be to seek to understand the desired outcomes of a complexity attack. To *impose or exploit complexity* is to take an action that increases an aspect of the complexity of the environment in a way that makes it more difficult for an adversary to make decisions or to operate, essentially shaping conditions to favor Blue. Thus, to conduct a *complexity attack* is to take an action that exploits characteristics of CAS to have a deliberate negative effect on the adversary. We can inform actions we take by considering complexity. Similarly, the CAS lens can help analysts see the extent to which a given system incorporates CAS principles and understand what parts of the CAS space are more or less represented within a single system or across a set of systems. Here, vignettes may help illustrate what complexity looks like (i.e., the independent variables to be manipulated) and what complexity achieves (i.e., the dependent variables to be influenced) in MDOs. Multidomain operational concepts are exemplified by recent Air Combat Command (ACC) and Training and Doctrine Command (TRADOC) MDO wargames.

There is a difference between the complexity in a system under attack and the complexity of an attack on a system. For example, an attack against a complex system might be very simple if we know just the right part of the system to attack to achieve a major effect. Alternatively,

---

[3] In fact, Air Force Research Laboratory (AFRL) is also concerned about the complexity of U.S. C2 and the ability of the adversary to conduct complexity attacks against the United States.

affecting a complex system might require a complex attack.[4] System complexity can be measured in a number of ways. These include quantifying how hard it is to describe a system, how hard it is to create the system, and to what degree the system is organized. However, the key consideration in the context of military C2 is impact. Measures of complexity are meaningful only insofar as they relate to—and predict—the desired outcomes of applying effects. If adopted by the Air Force for C2 use, the definition and measure of *complexity* should consider human physical and cognitive limits, magnified by the stress and pressure of responsibility, exacerbated by information uncertainty (e.g., environmental, organizational, human), and compounded by the nonlinearity of combat. In addition, we should consider ways in which intelligent automation could be developed and used to overcome human physical and cognitive limits for both defense and offense. In other words, both the human factors related to complexity and the decision aids that leverage artificial intelligence to help warfighters are critical factors in complex warfare evaluation.

The research questions for this effort are as follows:

- What does *complexity* mean in an Air Force context?

  - How is it defined?
  - Can it be defined in a way that makes it operationally relevant?

- How can complexity be employed in operational settings?

  - How can we model its impacts?
  - How can we measure it?

- How can this research on the nature of complexity inform science and technology (S&T) efforts with respect to complexity in warfare?

Research Approach

Our research approach consisted of four steps. First, we conducted a literature review on CAS, MDOs, and complex warfare (Chapter 2). The literature review helped define *complexity* in the warfighting domain and distinguish it from other concepts, such as fog of war, friction, paralysis, operational unpredictability, and deception. We reviewed the literature on complexity theory and how the imposition of complexity on adversaries shapes their actions. Key attributes of complexity and related metrics were obtained from the research.

Next, we developed four vignettes to explore what complexity might look like (i.e., the independent variables to manipulate) and the outcomes it may achieve (i.e., the dependent variables to measure). The focus on *complexity imposition* in our analysis relates to the

---

[4] A simple attack on a C2 node may be a kinetic strike, which could result in a very clear impact on a complex system. A complex attack example is introducing errors to C2 messages from that C2 node. How to evaluate the impact on adversary's decision processes and help select between the two alternatives are issues that may be aided by applying a CAS lens.

adversary's decisionmaking as it navigates the decision calculus to arrive at strategic or operational decisions. Chapter 3 presents the agile combat employment (ACE) and suppression of enemy air defense (SEAD) vignettes. We developed vignettes leveraging recent wargame work with TRADOC and ACC on MDOs and agile basing concepts considered by warfighters in the Pacific and European theaters. The project developed a framework based on the adversary's decision process to enable a systematic study of C2 complexity and quantification of return on investment from its imposition and associated costs.

Third, we developed a complexity *lens* based on the CAS research to characterize the multidomain operational environment and various types of attacks on the adversary's decisionmaking process. By looking through this lens at adversary C2 systems, opportunities to exploit complexity characteristics of the C2 systems in an attack are highlighted. The lens frames operational plans in terms of CAS characteristics and can aid thinking of planners toward shaping the adversary's decision calculus through a more effective selection and employment of effects. The complexity lens is then applied to the AFRL autonomy and C2 research portfolio to illustrate the application for evaluating science and technology (S&T) efforts that address aspects of complexity attributes.

Finally, based on the framing of complexity exploitation through the vignettes, we build a mathematical representation of the decisionmaking cycle and how different U.S. actions can create feedback loops and other complexity characteristics on the cycle. The model provides an example of how AFRL might simulate these attacks.

The report concludes with recommendations on how AFRL might move forward in (1) the study of complexity imposition, (2) the measurement of its effects, and (3) future applications of framing joint all-domain C2 warfighting in terms of creating complexity on the adversary's decision process (Chapter 4).

2. Complex Adaptive Systems

Creating a Complexity Lens

In this chapter, we discuss the literature on complex systems and how research on CAS pertains to warfare. The concepts of warfare presented are not new; rather, we show a framing of warfare in terms of complexity that allows planners, operators, and analysts to think about warfare through a lens of complexity. The lens provides a new vocabulary for describing aspects of war, a means to understand how U.S. systems may be vulnerable, and suggestions for how to gain advantage.

The literature on complex systems commonly distinguishes among simple, complicated, complex, and CAS. Something that is *complex* has many parts that are interconnected and interdependent and may create emergent behaviors. The system components are inanimate and do not adapt. In contrast, *complex adaptive systems* have components or agents that are animate and adapt. It is not always clear what causes behaviors, and causes are almost always multisource. The system can demonstrate emergence, such as through a chemical reaction (M. Mitchell, 2009; S. Mitchell, 2009). Volume II, the companion report to this document, expands on this discussion (Lingel et al., 2021). To understand complexity concepts as they apply in warfare, it is useful to think about warfare—and all of the people, institutions, equipment, terrain, and other relevant elements—collectively as *CAS*. This conception is useful because the behavior of each belligerent depends on the behavior of its adversary, as well as the details of its own state. Small changes can have large effects, as evidenced in the Battle of Midway, when Vice Admiral of Japan's fleet Chūichi Nagumo decided to change out the armament of his airplanes, leaving them targets to the U.S. dive bombers that attacked his carriers.

We conducted a review of complex systems literature in a broad range of disciplines, including complexity sciences, computer science, philosophy, and the natural, physical, and social sciences. In the course of the review, analysts identified a number of characteristics consistently related to CAS. We then trimmed the list of characteristics to reduce concept overlap and organized it into two bins: properties of complex systems and avenues for complexity imposition. Figure 2.1 shows the CAS lens.

Figure 2.1. Complex Adaptive Systems Mapping to Warfighting

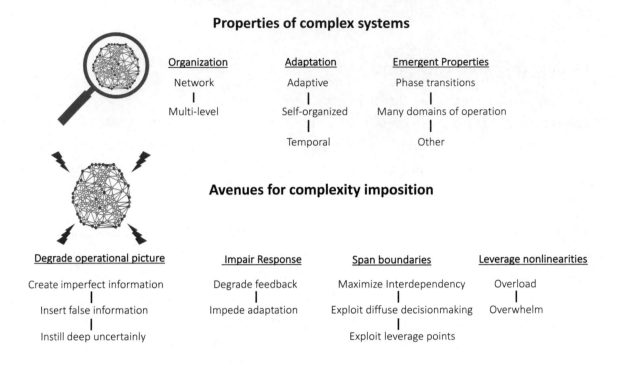

Properties of complex systems

Organization	Adaptation	Emergent Properties
Network	Adaptive	Phase transitions
Multi-level	Self-organized	Many domains of operation
	Temporal	Other

Avenues for complexity imposition

Degrade operational picture	Impair Response	Span boundaries	Leverage nonlinearities
Create imperfect information	Degrade feedback	Maximize Interdependency	Overload
Insert false information	Impede adaptation	Exploit diffuse decisionmaking	Overwhelm
Instill deep uncertainly		Exploit leverage points	

The sections that follow unpack each of the properties and avenues in the rubric, describing them abstractly and pointing out how they are relevant to military concerns.

Leveraging the Properties of Complex Adaptive Systems

A means of using complexity in warfare is to leverage the complexity of systems to create more-challenging conditions for an adversary for U.S. advantage. To *impose or exploit complexity* is to take an action that increases an aspect of the complexity of the environment in a way that makes it more difficult for an adversary to make decisions or operate or, in other words, to shape conditions in U.S. favor.

This section briefly describes the properties of CAS that may provide opportunities that the United States can leverage (and likely already does) in strategic competition and warfighting. Conceiving of warfare in these terms is the first step in viewing it through a CAS lens. As reflected in Figure 2.1, the CAS properties are divided into three categories: organization, adaptation, and emergent properties. In the sections that follow, we further describe each of the properties and actions listed in the figure, characterizing them abstractly before outlining how they might apply to warfighting.

It is important to note that there are many ways that the United States can be successful; this lens is designed to highlight the extent to which CAS principles are central to the design of strategic competition and warfighting plans and actions.

Organization

Central to the idea of a CAS is that it is organized as a *network* (Estrada, 2011). Specifically, many types of complex adaptive systems tend to organize as a hierarchy of nested nodes. To be *nested* means that systems and networks are placed within larger systems and networks. In a social system (such as a conflict), the nodes in the network may be individual people, organizations, or even pieces of equipment.

Many complex systems are organized into *multiple levels*, such as by rank, personnel, or operations. Different levels of a system have different characteristics, behaviors, and rules, and higher-level systems assume the rules and behaviors of those beneath them in addition to possessing characteristics of their own (Mazzocchi, 2008; National Academy of Sciences, 2019). Militaries are multilevel complex systems characterized by hierarchical command structures, which can be seen as nested, and there are horizontal connections at lower ranks among soldiers with interdependent responsibilities.

We can speak meaningfully of a U.S. response to an adversary even though both the U.S. military and the adversary military comprise complex arrangements of people, doctrine, communications, equipment, and the like. Similarly, we can talk about a headquarters unit or a flight crew without understanding the functions of each of the individuals who compose these units. Both U.S. and adversary headquarters and flight crews are examples of functional subsystems within the context of a larger complex system.

In summary, to understand warfare in terms of CAS and in particular the structural organization of the adversary's C2 system, it is instructive to characterize the extent that the adversary's C2 system is a network structure and how much of the system involves multilevel nested networks. These characteristics, discussed in the next section, present avenues for complexity imposition on the adversary's decisionmaking. Understanding the details of the adversary's C2 organization and processes is key to informing effective U.S. efforts.

Adaptation

Another essential property of CAS is that it adapts over time (Murphy, 2014). Agents within the system react to decisions made by other agents and in response to changing environmental conditions (Murphy, 2014). The result is self-organization in which a system's parts respond collectively to challenges without central coordination (Garfinkel, Shevtsov, and Guo, 2017). Military units are organized based on doctrine and past experience to achieve relevant ends within a context. The adversary's C2 organization can be expected to change in response to changes in many areas, including strategic priorities, U.S. tactics, technology and equipment, resourcing levels, past performance, and other drivers. Although some of these changes will be the result of top-down orders, others will be bottom-up shifts based directly on experience.

Thus, the decisions and changes that drive adaptive restructuring of CAS tend to come from all levels of the network—from the top to the bottom. This kind of adaptation can be particularly effective in a military context, when clear strategic guidance coming from the top and best operational practices coming from the bottom can be quickly identified and disseminated.

One implication of this form of adaptation is that it is *temporal*: It takes *time*. It may take time for lessons to be learned, structure to be adapted, and adaptation to propagate. Understanding an adversary's C2's speed of organization and adaptation can be critical to understanding how it will respond to change and stress.

In summary, the inability of an adversary to adapt to changes, how an adversary may or may not be self-organized, and the temporal aspect of its ability to adapt are important CAS characteristics that the United States may exploit. These characteristics, examined in the next section, afford avenues for complexity imposition on the adversary's decisionmaking.

Emergent Properties

A CAS tends to have properties that cannot be predicted by or understood in terms of the properties or behaviors of their constituent components. For example, one person cannot be *at war* (although this term tends to be used figuratively); instead, war is a collective activity requiring many people, sometimes working in concert with allies, to defeat shared enemies.

Phase transitions are a major emergent property of many CAS (Morris et al., 2019). Phase transitions consist of a qualitative change in which new systemic conditions emerge, such as new rules or new characteristics. Different phases operate by different rules. They can be precipitated by an event, such as the self-immolation that triggered the Arab Spring or the assassination of Franz Ferdinand that ignited World War I. When a phase transition occurs, rules, culture, and expectations change—for both physical and social systems. When someone says, "everything changed after 9/11," they are expressing the concept of a social phase transition.[5] Other examples of a phase transition—in a military context—would be a C2 system becoming overwhelmed or the transition from a hostile standoff to a pitched battle.

Domains of operation can be conceived as physical or, in the case of cyberspace, nonphysical places *in, from, and through* which capabilities can be employed to create desired effects within that domain or other domains. Operations are taken *in, from, or through* a domain with the aim of providing an actor with greater freedom of action to achieve its mission objectives in that or other domains or conversely to deny an adversary the freedom of action necessary to achieve its mission objectives. Domains of operation can usefully be thought of as CAS that emerge not as the result of the use of any one capability to create a particular effect but from multiple kinds of

---

[5] The boiling of water is an analogous example of a phase transition. The H_2O molecules in liquid phase each have a level of kinetic energy, but, because they are tightly packed together, this energy can only be manifested as vibrational and rotational motion. However, at 212° F (corresponding to a certain average kinetic energy level at a certain atmospheric pressure), the molecules no longer constrain one another and can produce translational motion, resulting in the dramatic transition from liquid to gas phase, in which the substance takes on radically different properties. If we observed a single molecule at different energy levels, we would see no evidence of a transition as it gained energy: Boiling is a collective activity that has meaning only in the context of a large collection of molecules.

operations taken in, from, and through that CAS over time. To illustrate, contemporary U.S. military doctrine recognizes five operational domains (i.e., domains of warfare): land, maritime, air, space, and cyberspace. The land and maritime domains are, of course, the oldest warfighting domains; the air domain did not fully come into its own until World War II.[6] The development and increasing use of satellites for early warning, intelligence, and communications during the Cold War marked the emergence of space as an operational domain of critical importance to strategic nuclear operations specifically and strategic stability more generally. But it was only in the post–Cold War years, and more specifically in the wake of the first Gulf War and subsequent U.S.-led NATO operations in Kosovo in the late 1990s, that space came to be recognized as a domain essential to enhancing the real-time operational effectiveness of advanced nonnuclear land, maritime, and air operations (Watts, 2001). Finally, although computing and information capabilities had been used since the mid-1990s to engage in a range of offensive and defensive cyber operations, cyberspace only began to be widely acknowledged as a domain that could be exploited by adversaries to threaten essential U.S. interests following a major compromise of the U.S. Department of Defense's classified military computer networks in 2008. Just two years later, cyberspace was formally recognized in U.S. military doctrine as an operational domain (Lynn, 2010).

Other useful groupings include basic warfighting functions (C2, movement and maneuver, intelligence, fires, sustainment, and protection), the division between officer and enlisted, and the division of the Air Force into wings, groups, squadrons, and flights. Although these groupings may be somewhat arbitrary, they generally derive from the principle that people and equipment with tightly integrated functions should be managed together and that each commander should have oversight of a (humanly) manageable number of subelements. Although the precise structure of the Air Force may be a product of historical development, the functional demands under which it operates and the capacities of the people who staff it were bound to produce an organization with the same general character.

To summarize, possible emergent properties exhibited by the adversary's C2 system may be phase transitions and domains of operation. It may help to comprehend the extent to which Red's C2 system has phase transitions, what they are, and how the system spans domains of operations. Attacking at the seams between two domains is one possible Blue attack vector and aligns with thinking on MDOs. These CAS characteristics, as we discuss in the next section, provide avenues for complexity imposition on the adversary's decisionmaking.

---

[6] Despite the fact that observation balloons were used by both federal and Confederate forces in the American Civil War, the air domain was not generally recognized as a primary operational domain until fighter and subsequently bomber aircraft came into widespread and effective use in World War I and more emphatically World War II.

Using Complexity to Attack the Adversary's Decision Steps

By viewing conflict as CAS, the complex features of an environment can be harnessed to create more-challenging conditions for an adversary or can be used to attack directly. We describe these as *complexity imposition* and *complexity attack*. To impose complexity is to take an action that increases an aspect of the complexity of the environment in a way that makes it more difficult for an adversary to make decisions or to operate, essentially shaping conditions to favor Blue. A military deception campaign nested in an operational scheme of a maneuver plan is an example of complexity imposition. One example would be deception campaign paired with an ACE movement plan. To conduct a complexity attack is to take an action that exploits characteristics of CAS to have a deliberate negative effect on the adversary.

We identified four general categories of actions that gain their utility directly from the CAS nature of the system. These actions include degrading the operational picture, impairing Red response, spanning organizational boundaries, and exploiting nonlinearities. Examples are interwoven in the following section to illustrate these categories.

Degrading the Operational Picture

In most CAS, the component parts of the system (whether people or pieces of equipment) have neither perfect information about their adversary and environment nor perfect reasoning ability to process the information that they do have, given the constraints of time, memory, and the like. Operational efficiency requires that good information be collected and processed by the system. This avenue of complexity imposition is directed at the information environment.[7]

Because decisions are being made at all levels of an organization, and decisions depend on observed and anticipated adversary actions (which, in turn, depend on adversary observation and anticipation of Blue actions), CAS decisionmakers are always operating with limited information and limited cognitive processing capacity. Although each side strives to know as much as possible—and to think as clearly as possible—about its opponent, the CAS perspective acknowledges that such limitations are inherent and looks for ways to minimize the impact of these limitations on Blue and maximize the impact on Red.

Information degradation can take several forms. First, Blue can create imperfect information and ensure that Red knows less than Blue does about what is happening. This effort takes the form both of maximizing Blue access to information and of minimizing Red access to information. Second, Blue can feed false information to Red—leading Red to make bad decisions or to make slower decisions as it works to separate good from bad information. The history of armed conflict offers many examples of successful military deception (MILDEC) that can be productively analyzed through the lens of CAS. Although the first two classes of

---

[7] CAS literature refers to this avenue as *bounded rationality*, which is a property specific to complex adaptive human systems (Simon, 1972).

degrading the operational picture focus on denying information or feeding false information to the adversary, they take the form of maximizing the adversary's *known unknowns*—the adversaries know what they want to know but either cannot discover it or get it wrong. A third form of deception through information degradation is the creation of deep uncertainty: In a situation of deep uncertainty, there are unknown unknowns (Marchau et al., 2019). This means that the nature of the problem or the objectives of a mission cannot be understood (Lempert, 2002; Lempert et al., 2013; Lempert, Popper, and Bankes, 2003). "Wicked problems" have this character (Churchman, 1967; Rittel and Webber, 1973). Blue can gain advantage by developing a clearer understanding of its own goals and objectives than Red and by developing a relatively better understanding of Red's goals and objectives than Red has of Blue's.

Blue can also exploit deep uncertainty by creating situations in which Red does not understand how the system works. This may lead Red to focus its efforts (intelligence or otherwise) in areas that are not the most useful or to take actions that do not lead toward its desired outcomes. However, a caveat is needed here because Blue does not always want Red to make bad decisions. In some cases, bad decisions on the part of Red can lead to outcomes that are worse for both parties. In a deep uncertainty situation, Blue needs to remain aware of when flawed Red decisions are likely to produce benefits for Blue.

Impairing Red Response

The CAS lens provides perspective on how forces adapt to changing circumstances and to changing adversaries (Kwakkel, Haasnoot, and Walker, 2015). As just mentioned, the self-organized adaptation of human CAS involves observations and decisions that are made at many different levels of the organization. Generally, it is advantageous for Blue forces to be quicker and more adaptable than Red. This can, however, prove challenging because the U.S. military most recently has been larger and more mature than its adversaries, with more-developed doctrine and clearer procedures. Although these properties provide real advantages in familiar environments, they also create a degree of inertia, making it harder for the organization to change and adapt to new situations. However, the United States is not currently equipped with doctrine and tactics, techniques, and procedures for MDOs.

Feedback is an essential feature of all CAS and is the first critical step in adaptation. Generally, gathering feedback is the process of observing the outcomes of actions and using these outcomes to inform future ones (Brinsmead and Hooker, 2011; Grösser, 2017). This is a primary function of military intelligence. The U.S. military needs to establish systems for making use of feedback and adapting its structure, tactics, goals, and behavior to meet the challenges it faces (Diehl and Sterman, 1995; Rickles, Hawe, and Shiell, 2007). The CAS lens draws attention to the advantage that can be gained by doing this more quickly and effectively than Red forces can do the same thing.

The United States will benefit from understanding the adversary's C2 ability to adapt in warfighting. For example, what is the adversary's C2's reaction to the environment or to alternative U.S. attacks from different domains? And does the adversary respond in a predictable way? Knowing the extent to which the changes are determined centrally—or decentralized, with subcomponents reacting and adapting independently—further informs Blue courses of action.

Spanning Organization Boundaries

Conceiving of both Blue and Red militaries as networks suggests various methods that Blue might use to gain advantage. The nested structure of these networks means that not all nodes have equal value and invites analysis of which interventions might be most effective in preserving Blue capabilities and disrupting Red capabilities (Ruiz-Martin, Paredes, and Wainer, 2015).

Interdependencies create the opportunity for cascading impacts. A disruption in one node may create disruptions in a string of nodes that are dependent on that node. This downstream disruption, in combination with disruptions from other upstream nodes, can cause major problems for downstream nodes. This disruption can both disrupt critical functions and force Red to devote precious time to the examination of its own position, leading to a slower and less-effective response.

Diffuse decisionmaking is often a defining characteristic of self-organization in human-based CAS. Systems tend to evolve to a point where decisions are made at all levels throughout the organization, with each level of command making decisions based on the authorities delegated to it. On the one hand, this kind of decisionmaking structure can be extremely robust, allowing part of the organization to operate, even when important nodes are disrupted. On the other hand, it can complicate decisions involving multiple decisionmakers who need to coordinate on a common action. An example is a mission using cyber and kinetic effects in which cyber and kinetic resources are planned and resourced through two different C2 organizations; this can cause delays and degradation in resourcing and effectively employing kinetic and nonkinetic capabilities in an integrated operation.

A major implication of the hierarchal structure of military networks is that advantage can be gained by forcing the adversary into actions that require the coordination of nodes under different command. A Blue action that requires a Red response from a cohesive group that has trained together and answers to a single commander is not as powerful as an action that requires a Red response from several groups located in different places and different branches of the command hierarchy. Response in such a situation will rely on communication that is less efficient and decisionmaking that involves either the development of consensus or additional levels of command. In either case, the boundary-spanning response is likely to be slower and less effective.[8] An example of exploiting the complexity of Red's C2 to the advantage of Blue is a

---

[8] Conversely, Blue should anticipate such actions on the part of Red and should implement boundary-spanning mechanisms to handle these actions. These might take the form of new doctrine, the establishment of staff liaisons, the conduct of joint training exercises, and the like.

Blue attack on the Red military network that spans organizational boundaries. In a later section, we will discuss network analysis on an adversary's decision processes.

CAS network structures can also be exploited by finding **leverage points** within them (Meadows, 1999; Russell and Smorodinskaya, 2018). The inherently uneven nature of both C2 networks and the network of unit interdependencies means that there are bound to be some parts of the C2 network where a relatively small disruption can have a large effect (Hofstetter, 2019). For example, Red may have commanders or command systems that are critical to its decisionmaking: These are high-value targets that are susceptible to disruption. In identifying such leverage points, it is crucial to understand the nature of adversary C2. In addition to formal chain of command, peer-to-peer communications and information systems may provide critical links within adversary C2, which do not have the same structure as the formal chain of command. These links may provide resilience in the face of chain-of-command disruption but also may provide vulnerabilities that are not apparent when examining chain of command alone.

Operational interdependency suggests a different network that may also present leverage points. By tracing the parts of the adversary C2 that depend on one another, it may be possible to identify critical bottlenecks. As a simple example, Red may have a single fuel depot located at a port that serves land and air operations in a battle space. Even if the Army and Air Force use different fuels and manage them under different commands, the physical colocation produces a common network node whose functionality is essential to the conduct of multiple operations across two services: This would be an extremely high-value node. Although this example is rather obvious (fuel depots are a classic point of attack), the network concept provides a general method for identifying and ranking high-value nodes.

The interaction of this operational dependency network with the command-structure network provides opportunities to choose actions that require response from interdependent units that are closer in functional relationship than they are in command relationship. Response in such situations will incur overhead as units work to coordinate their actions and develop consensus without direct command oversight. In cases where the need for such coordination is common or expected, boundary-spanning mechanisms can be expected to have evolved to facilitate such coordination. However, actions that require a response from groups that do not commonly work together may lead to delay, miscommunication, and other dysfunctional responses that work to the advantage of Blue.

Leveraging Nonlinearities

Nonlinearities are one common characteristic of CAS, where small changes in input can produce large (or even discontinuous) changes in output. The phase changes discussed in the emergent properties section of this chapter are examples of such nonlinearities (Richardson, Paxton, and Kuznetsov, 2017).

Overloading a Red node in the network is an example of leveraging nonlinearities. A commander might perform well under a light load and continue to perform well as the load increases, making only the occasional omission or error, with errors increasing in proportion to the workload. However, at some point, the information-processing and communication capability of this person becomes overloaded, resulting in many more things being dropped. The falloff in performance beyond this point is likely to be quite steep because the commander is unable to handle or process additional information. The same logic applies to the overloading of any C2 system. Overloading, as described here, is an example of a phase change that is highlighted by the CAS lens.

Another nonlinearity is the phenomenon of **overwhelming**. When Blue attacks and Red holds a material advantage, Red is likely to fight and win—possibly with light losses. When the two sides are evenly matched, the outcome is uncertain, but losses are likely to be heavy on both sides. When Red is at a slight disadvantage, Blue is more likely to win, and Red may take heavy casualties. However, when Red is at a substantial disadvantage, it is likely to retreat (or surrender) immediately. In this situation, Red is overwhelmed by superior force and is not able to put up a viable defense. The difference between the case where Blue has a small advantage and the case where Blue has an overwhelming advantage may be minor depending on how Red views its prospects, but the difference in outcomes is quite large.

The CAS-related opportunities for Blue advantage here are Blue's ability to overload the adversary decisionmaking process, which slows down that process or causes the adversary to make poor decisions, thus creating decision advantage for Blue. Another opportunity is when Red C2 is overwhelmed by Blue's force (which need not be kinetic), thereby causing Red to halt aggression.

Applying a Complexity Lens

This chapter described the CAS properties that Blue forces can leverage for military advantage and the means of complexity exploitation enabled by these properties. We describe this thinking of Air Force operations in terms of the CAS properties as applying a CAS lens to warfighting. None of the actions discussed here are new, but we find that the CAS lens is a useful way of identifying alternative courses of action and seeing their utility in modern complex operations. The CAS lens may highlight combinations of CAS properties that may have amplified effects to U.S. advantage. Furthermore, the CAS framing of warfare focuses on the adversary's decisionmaking processes or systems and imposing complexity by drawing on capabilities across the domains of warfare—land, sea, and air, as well as cyberspace and space— in a manner designed to exploit vulnerabilities in the adversary's C2 because of the CAS nature of decision processes. For a guide on using the CAS lens for considering S&T research that explores exploiting complexity for military advantage, see the second volume of this report.

In the next chapter, we present an example operational-level vignette to better understand avenues of complexity imposition on the adversary's decisionmaking. The example is illustrative and provides a backbone with which we can later describe a mathematical representation of these concepts.

3. Application of the Complex Adaptive System Lens to a Mission Vignette

In this chapter, we develop the concept of complexity imposition as a tool that can be used to shape outcomes in specific operational contexts. We explore the idea that, by imposing increasing levels of complexity on adversary decisionmakers, it is possible to shape their behavior in predictable and beneficial ways. As we described in Chapter 2, complexity can be understood as a set of attributes that are present in most CAS and whose state can be influenced by manipulating a particular set of system characteristics. This focus on the manipulation of complex systems offers a way to identify avenues through which a complex system can be influenced through external attacks, which we term *complexity imposition* or *exploitation*. These attacks shape the types of complexity that decisionmakers must contend with when orienting and reacting to an operational vignette. Accordingly, the target of *complexity exploitation* in our analysis is an adversary's decisionmakers because they navigate the decision calculus to arrive at strategic or operational decisions.

In the next sections, we first provide a framework for understanding the nature of complexity attacks and then present analytical strategies that can be used to understand the operational impacts of these efforts. Our goal, ultimately, is to quantify the effects of complexity imposition on decision processes in the warfighting domain.

Modeling an Adversary's Decision Calculus

We begin our analysis by laying out an adversary's decision calculus as it applies to a specific operational context. This represents the base state of adversarial decisionmaking and Blue operations, both traditional and complexity focused, that attempt to shape the outcome of this decision flow to achieve a desired set of operational effects. In some cases, the outcome might be more robust deterrence as Red sees that the likelihood of arriving at a desired end state is diminished. In other cases, the outcome might be an inability of Red to act quickly as necessary information is denied and risk-reward trade-offs are more complex.

We focused on decisionmaking as the target of these attacks because, as we explored possible analytical frames, this provides the clearest articulation of ways that complexity exerts its influence in operational practice. Early in our study, for example, we experimented with ways to model complexity's impacts on the observe, orient, decide, act (OODA) loop, which frames tactical military action as being made up of a cyclic step of processes that yield decisive advantage if they can be completed faster and more accurately than an adversary who is completing a parallel OODA loop process of its own.[9] We found that increased complexity affected every stage of the OODA loop process, not simply the decide stage. Unclear signals, concealment, and decoys can confuse or slow down the observe phase, while increasingly complex environments create challenges during the orient phase. As we analyzed examples at each stage of the OODA loop and considered the factors that might make it relatively faster or slower to complete the loop, we identified the load imposed on decisionmakers at each step—what to observe, how to make sense of signals, how to convert a situational picture into a plan of action, and how to carry it out—that were the key areas of impact. Following this analysis, we modeled the impact of complexity on decisionmaking processes, independent of their position within the OODA loop or other frameworks.

To demonstrate the conceptual approach and provide a relevant warfighting example to ground our discussion, we apply this framework to an operational vignette based on the Air Force concept of ACE.[10] In this context, the target of the complexity attack is an adversary's strategic decision calculus—the adversary's assessment of whether an offensive first strike is likely to enable it to deny Blue's freedom of maneuver, which would allow it to achieve its strategic goals.

Additionally, we apply this framework to tactical decisionmaking in a SEAD vignette. This vignette considers the decision framework that an adversary must navigate when attempting to maintain robust air defenses in the face of a Blue air campaign. Although it is at a more tactical level than the ACE vignette, it is a useful addition because it clearly and intuitively conveys the trade-offs and implications of complexity imposition.

As the notional decision flow in Figure 3.1 depicts, an adversary is weighing the predicted outcome of a first strike against U.S. bases. Following an increase in tensions that bring kinetic options into consideration, an adversary must then assess both U.S. posture and vulnerability as well as its own capabilities to arrive at a predicted operational outcome of a first strike. This expected outcome is the basis of a risk-reward calculus that, if it is viewed as sufficiently favorable, will result in a decision to launch a first strike. This is the strategic-level operational threat that the ACE concept is intended to address. Diminishing U.S. posture vulnerability, for example, presents a more robust deterrence to attack.

---

[9] The OODA loop is often referred to as Boyd's OODA loop.

[10] *Agile combat employment* refers to operating in an adaptive manner from a fluid set of basing locations to render the adversary's targeting of U.S. forward forces more complex.

Figure 3.1. Adversary Decision Calculus for Strategic First Strike

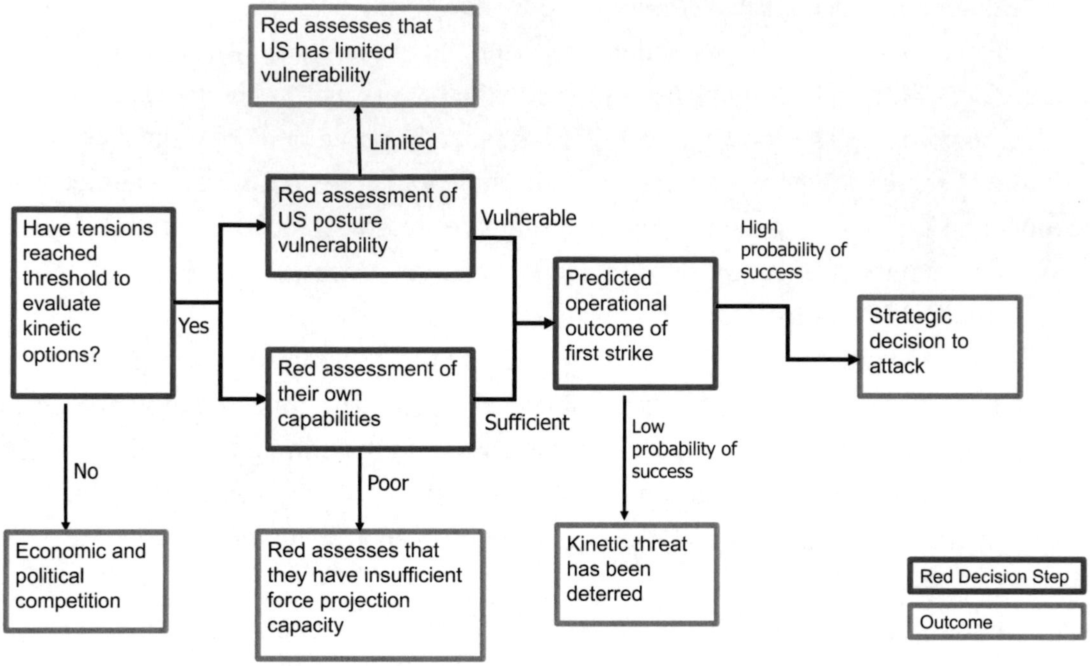

Applying Complexity Concepts to Deterrence

To demonstrate the application of the CAS framework, we first consider the range of tools available to make U.S. at-risk bases more robust and survivable. One of the tenets of the ACE concept is that U.S. bases can lower their vulnerability (both the likelihood of being attacked and the likely damage incurred) to attack in various ways that limit the potential damage to air operations caused by adversary attacks. Active defenses, for example, can intercept incoming missiles and deter attacks, while shelters can lower their ability to damage U.S. aircraft. Such tactics as dispersing aircraft within a base achieve a similar effect by reducing the damage that an individual missile can inflict, while base resiliency measures, such as runway-repair capabilities and auxiliary fuel bladders, can enable the rapid resumption of operations following an attack.

Viewed in terms of Red's decision flow, such measures lower the perceived assessment of U.S. posture vulnerability once Red becomes aware of them (Figure 3.2).

Figure 3.2. The Impact of Base Resiliency Measures on an Adversary's Decision Calculus

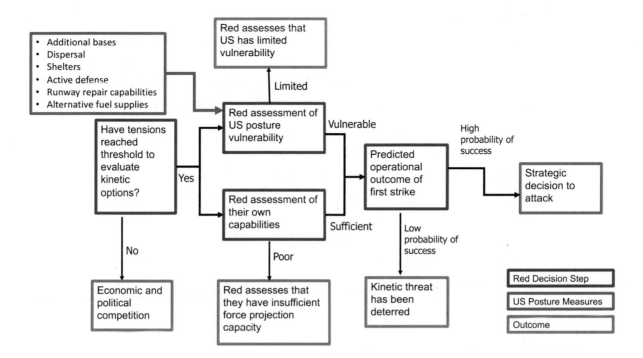

As Figure 3.2 illustrates, the United States can influence an adversary's decision calculus by changing the posture vulnerability assessment in favor of the United States. Viewed at a strategic level, the operational impact of these measures is to achieve deterrence by reducing the likelihood that an adversary will decide to attack.

One significant implication of this framing is that the range of U.S. options to increase base survivability does not reshape an adversary's decision process. That is, they do not introduce new variables or considerations into the decision calculus. Rather, they affect the ultimate risk-reward trade-off by making the likely outcome of a base attack less beneficial to an adversary. Thus, the overall decision flow is unchanged, but the parameters associated with it are altered to be more favorable for U.S. interests. In contrast, the impacts of complexity imposition described in the next section are primarily achieved by forcing an adversary to consider a range of new decisions whose outcomes are beneficial to U.S. interests and that impede its ability to move along its strategic decision tree.

Complexity Attacks: Reshaping Adversary's Decision Flows

The example just outlined illustrates that deterrence does not increase complexity. Next, we turn to modeling the operational impacts of ACE strategies that leverage complexity—to deny the adversary an accurate operational picture and effective targeting (by employing complexity). Instead of simply trying to protect U.S. assets and make them more survivable, as the measures described in the previous section did, here we consider the roles of deception and information denial in shaping adversary behavior. If, for example, the United States were to conceal the presence of aircraft with camouflage or move them to new bases more quickly than enemy intelligence, surveillance, and reconnaissance (ISR) could locate them, the ability to reliably target U.S. aircraft would be called into doubt. Likewise, dazzling, jamming, and other counter-ISR activities would make the task of finding, fixing, and targeting U.S. forces a much more difficult and fluid problem. Additional U.S. efforts might further complicate decisionmaking by corrupting Red data (through a cyberattack) about the potential efficacy of Red's own systems or might degrade decision aids (through psychological operations or cyber means) so that they cannot provide Red decisionmakers timely assessments of different courses of action.

The factors underlying these strategies for complexity imposition can be identified by matching each notional approach with the elements identified in the complexity lens in Figure 3.3. The example ACE vignette is by no means all inclusive, but, instead, it is illustrative of the complexity lens thinking. Cyber and counter-ISR actions are represented. Extending the adversary's decision processes to kinetic weapons arriving at an ACE base would provide a more complete picture of adversary MDOs—for example, satellites and airborne ISR that help with kinetic targeting that may be supported with electronic warfare.

Figure 3.3. Framing Elements of the Agile Combat Employment Concept of Operations in Terms of Complexity Imposition

Avenues for complexity exploitation		Military deception	Counter-ISR	Corrupted data	Degraded decision-aid	Kinetic weapon
Organizational properties	Network		✓		✓	✓
	Multi-level				✓	✓
Degrade operational picture	Create incomplete information	✓	✓	✓		✓
	Insert false information	✓		✓		
	Instill deep uncertainty			✓	✓	
Span boundaries	Exploit leverage points				✓	✓
	Exploit diffuse decision making	✓		✓	✓	
	Maximize interdependencies			✓		✓
Impair response	Degrade feedback		✓			✓
	Impede adaptation			✓		✓
Leverage non-linearities	Overload	✓	✓			
	Overwhelm	✓	✓			

By modeling the impacts of these strategies for complexity exploitation in terms of the adversary's decision flow, we observe that these concepts of operations (CONOPs) have outcomes that are very different from those that we previously discussed: Complexity exploitation introduces new decisions—and calculations—that adversary decisionmakers must navigate. By jamming adversary ISR collection and employing decoys to sow confusion about the true disposition of its forces, the United States does not simply lower the efficacy of Red attacks. It also introduces a new set of questions that adversary decisionmakers must consider. Given the imperfect data available, an adversary must now assess how trustworthy data are and whether to collect them, and it must determine whether imagery results are false positives or false negatives that will hinder accurate targeting. A notional example of this more-complex decision process is shown in Figure 3.4.

Figure 3.4. Impacts of Complexity Imposition on Adversary Decisionmaking

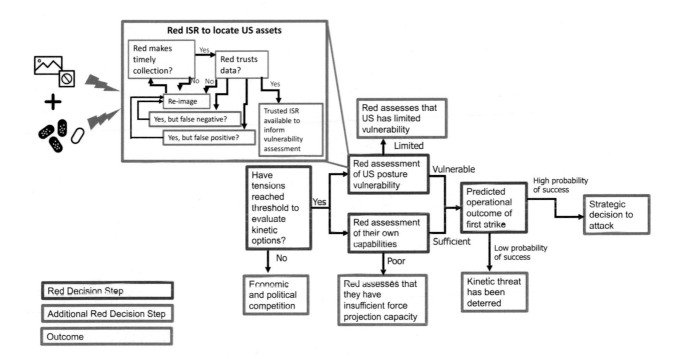

Although the green box in Figure 3.4 exists regardless of any Blue deception or counter-ISR efforts, the degree of uncertainty would increase with these complexity attacks. By imposing complexity here, we are introducing new decisions in this example. One can imagine that these Blue attacks would in fact increase uncertainty in each of the Red boxes. Although a clean distinction about what constitutes a complexity attack may be difficult to make, the application of the CAS lens (or framework) to military planning remains constructive.

This diagram demonstrates the imposition of additional decisions and operational considerations caused by the use of military deception (MILDEC) and ISR denial that deny an adversary a clear picture of U.S. force laydown. These measures make it more difficult for an adversary to advance through its decisionmaking to the point where it can definitively assess U.S. posture vulnerability. Instead, uncertainty and information denial leads to feedback loops in which additional imagery is needed. Moreover, the existence of false positive and false negative assessments may also reshape an adversary's decision calculus by sowing confusion. False positives, in which an adversary double counts transiting aircraft or is fooled by decoys, will cause decisionmakers to overstate the actual number of U.S. assets in the theater, causing them to lower their assessment of U.S. vulnerability or make a location appear a more lucrative target. False negatives, conversely, may result in lower-than-actual counts of U.S. forces, which creates an appearance of a smaller footprint, lowering the actual vulnerability of the overlooked forces because the probability of an attack would be lowered. Although we have included only a single type of example here, there are numerous possible avenues for complexity exploitation that can target all areas of an adversary's decisionmaking, as Figure 3.3 indicated.

Viewed in these terms, the difference between complexity exploitation and more-traditional actions is that a complexity-based response reshapes an adversary's decision calculus and takes advantage of the CAS's characteristic of the adversary's decision processes. There may be cases in which a given course of action exhibits both types of impact. This method is valuable because it provides a useful definitional frame that highlights the unique aspects of complexity-based attacks and allows the impacts of different types of Blue actions to be directly compared with one another in terms of their ultimate effects on an adversary's decisionmaking. And these actions can span from ISR, cyber, kinetic, and passive measures.

Given accurate information about the transition probabilities associated with each decision step or state, it is possible to compute the probability of arriving in a newly created (by complexity imposition) end point in the decision process or, in mathematical parlance, an absorbing state, at any point in time. This approach allows the impact of potential complexity attacks to be modeled by quantifying the success of false information and ISR-denial campaigns in sidetracking from the final decision to launch. This approach is developed in greater length in Volume II of this report, where we model the decision states as a Markov chain and then explore the implications of representing the transition states in matrix form, including in-depth discussions of the mathematical features of this model.

The *topology of the decision process* is the way in which decision steps and complexity exploitation actions are arranged or interrelated. Thinking about it this way allows us to construct the problem in a network form.[11] Viewed this way, several topological elements are indicative of the added levels of decisionmaking complexity that have been imposed on the adversary.

Introduced to the decision process are additional terminal states, which represent intermediate *dead ends* that impede adversarial decisionmaking by acting as absorbing states where decision chains can terminate. In the context of our example, these terminal states are such cases as false negatives in ISR collection that represent alternative potential end states of decision trees.

---

[11] It appears likely that, when U.S. planners have a more detailed understanding of the adversary's C2 processes, the planners can then have a more detailed mapping of the C2 network and a more tailored way of exploiting its complexity.

A second structural feature visible within the graph is the presence of feedback loops, which appear as cyclic paths within the decision tree. In contrast to the linear decision chains that structure our base-case example, these feedback loops suggest that a series of decisions may need to be made multiple times before a transition to a new pathway is achieved. In cases where ISR re-collection is required to determine the posture and thus the vulnerability of U.S. forces, these collection efforts might be ongoing, in which the outcome of the assessments is iteratively refined over time. The presence of these loops within the decision tree suggests that decisionmaking may be slower, because feedback loops can potentially make the decision pathways infinitely long.

Applying Complexity Concepts to a Suppression of Enemy Air Defenses Vignette

To clarify the presentation of our conceptual model and demonstrate its applicability to other operational vignettes, we have also applied our framework to a SEAD vignette. Understanding that the full operational flow associated with modern integrated air defense systems (IADSs) is quite complex, we consider a simplified model that highlights some of the main benefits of our modeling strategy.

Although our analysis of the ACE vignette in the previous section focused on adversarial decision flows, we begin our analysis here by focusing on tactical communications and unit-level dependencies rather than simply decisionmaking (see Figure 3.5).

Figure 3.5. Flow of Tactical Air Defense Response

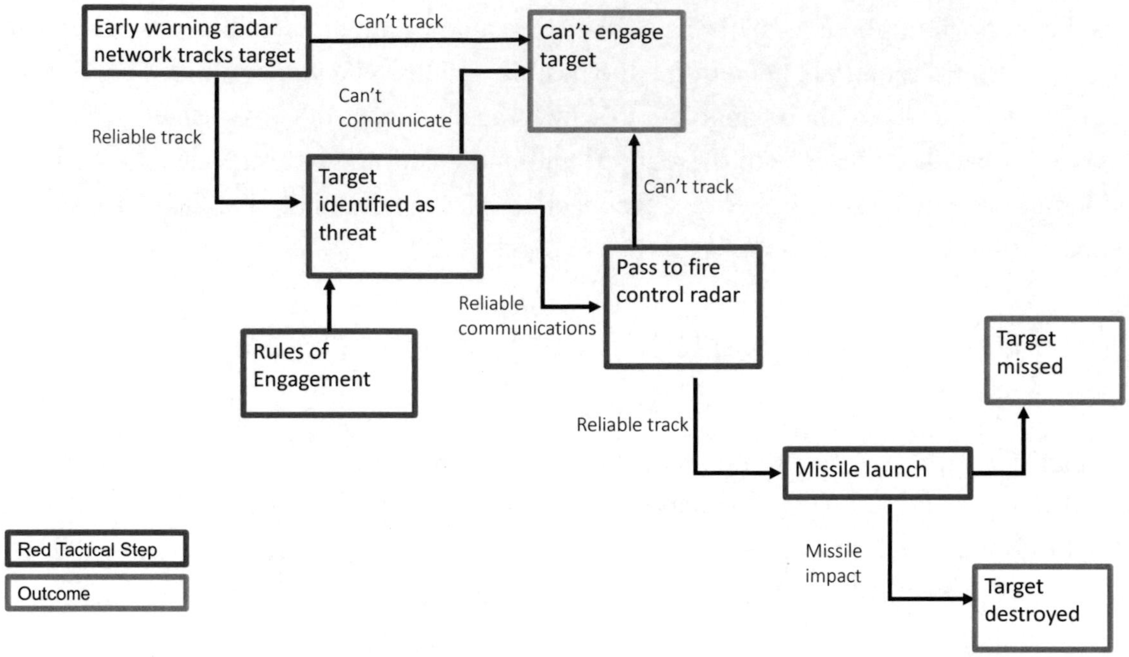

As in the ACE example above, there are a set of response options that deny enemy capabilities by lowering transition probabilities associated with different outcomes without fundamentally introducing new sets of decisions. As Figure 3.6 demonstrates, use of such technologies as radar jamming to prevent missile targeting or the deployment of chaff or flares to evade an incoming missile can be easily modeled in this framework.

Figure 3.6. Use of Defensive Measures to Influence Flow of Tactical Air Defense Response, Part 1

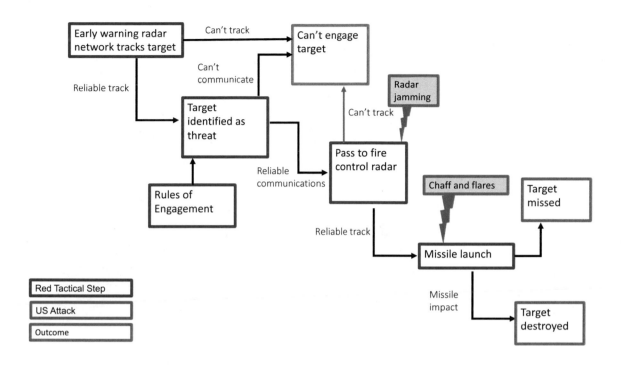

As this figure shows, these Blue measures succeed by negatively affecting an adversary's ability to move along its tactical chain of events required to destroy Blue aircraft. If a radar is effectively jammed, it cannot be used to guide missile launch, but, instead, the capability flow is temporarily shunted to an absorbing state in which a target cannot be engaged. After a missile is launched, its probability of kill can be negatively affected by deploying countermeasures that decrease the probability that the missile kills the target.

In addition to these defensive measures, there are also classes of responses that require adversary decisionmakers to consider a new range of possibilities and trade-offs. One example of this are efforts to influence the risk calculus of enemy air defense operators. Since the advent of high-speed antiradiation missiles during the Vietnam War, air defense operators have had to balance the need to operate radars to engage targets with the fear that doing so puts them at risk of attack. In some cases, the perception of risk might cause a surface-to-air missile (SAM) battery to move to a new location, making it unable to operate until it has set up at the new location. In other cases, crews might simply turn off emitting radars to hide their presence, an option that comes at the cost of being unable to track or engage incoming aircraft. Either response is beneficial to attacking aircraft because, in each case, the SAM batteries have temporarily removed themselves as immediate threats (Figure 3.7).

Figure 3.7. Use of Defensive Measures to Influence Flow of Tactical Air Defense Response, Part 2

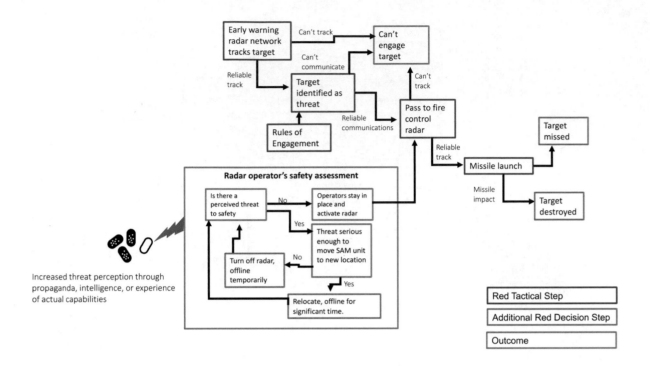

In our analysis, the manipulation of this threat perception can achieve an operational outcome that can be directly compared with such physical counter-ISR responses as jamming and chaff dispersal by measuring the degree to which they are able to steer the operational flow to end states other than successful missile engagements. Additionally, given the short time frames in which IADSs may be able to disrupt the initial phase of air operations, it may be highly beneficial to induce SAM crews to stay in the threat-perception feedback loop of moving and not emitting while the air campaign unfolds, unimpeded above them.

This counter- and countermeasure series of actions is well-known in the military. An example of a logical next step for the adversary here is to create passive means of detecting and tracking aircraft that removes the threat perception feedback loop above. We simply are presenting the counter- and countermeasure game here in terms of complexity.

Complexity Lens and Multidomain Operations

One advantage of framing competition and warfare with a complexity lens is that the impacts delivered by different operational domains can be analyzed in a common context. By modeling the range of available Blue effects that can be delivered through MDOs, the impacts of different combinations of effects can be readily identified for examination. Air operations can provide jamming or last-minute countermeasures, but equivalent effects might be achievable through cyber or space-based attacks that inhibit radar operation or effective communication. Likewise, when modeling the impact of threat perceptions on radar operators, the kinetic threat could be delivered through air-launched antiradiation missiles or from ground-based fires. By bringing together these different domains and ranges of operational effects into a common analytical framework, it is possible to directly compare their operational impacts.

Additional Considerations

Although not considered in this preliminary analysis, there are additional factors that should be considered to develop a more complete understanding of the operational utility of complexity.

One factor is the situations in which the imposition of complexity is not beneficial. The idea of the limits of complexity were acknowledged in a 2018 speech by then–Secretary of Defense James Mattis, who observed that the United States must be "strategically predictable for our allies and operationally unpredictable for any adversary" (U.S. Department of Defense, 2018). In operational practice, such deterrence scenarios as the Cold War mutually assured destruction function precisely because the game theoretic outcomes are clear to all players. Deterrence is achieved because the trade-offs are simple and complexity would only serve to undermine the balance that the doctrine achieved.

A second consideration that must be more clearly understood are the complexity costs that Blue imposes on itself as it undertakes to impose complexity on an adversary. Although complexity exploitation on an adversary might be operationally beneficial, it comes at a cost of C2 and coordination among forces and organizations. Weighing the relative cost on Blue against the effects imposed on Red will help to more clearly inform the trade-offs between strategies of complexity exploitation and those based on more-direct methods.

Furthermore, it may be that this trade-off between costs of implementation versus costs imposed will open a new domain for strategic advantage based on decision superiority. If one side in a conflict can exploit complexity at lower cost while better absorbing the complexity imposed on its decisionmakers by an adversary, they will hold an advantage on the battlefield.

Lastly, we must consider that complexity in warfare is not new and that numerous strategies already exist to manage its effects on military decisionmakers. Doctrine removes complexity from decisionmaking by providing clear guidance for what to do in different vignettes. Effective training likewise enables leaders to learn to operate effectively in complex situations so that they are not overwhelmed during actual combat. Even in the absence of these formal complexity-mitigation strategies, human cognition routinely relies on heuristics, analogies, and shortcuts—some beneficial, some inaccurate—that enable decisions to be made rapidly in vignettes that cannot be fully understood.

This report represents a first step at applying a CAS lens to warfare and, in doing so, notes areas for further exploration and analysis.

4. Conclusions and Recommendations

A key concern for the Air Force is the ability to leverage MDOs to U.S. advantage in both competition and warfighting. Multidomain actions are viewed as imposing complexity on the adversary's decision process, but a clear understanding of how and what the benefits are from this complexity imposition is lacking. In this report, we aim to enhance understanding of the nature and value of complexity as a weapon—that is, a complexity attack.

Our research is based on the hypothesis that strategic competition and multidomain warfare create complex environments for an adversary's decisionmakers and set conditions for U.S. forces to benefit from characteristics associated with complexity. If that is the case, a clear understanding of how complexity can be imposed and what value it brings to operational settings is needed to evaluate current and future capabilities and CONOPs. To ground thinking on multidomain warfare creating complexity for decisionmakers, we set out to provide a complexity lens from which to view great-power competition and warfighting. This complexity lens is grounded in CAS research.

Complexity in warfare is also not clearly defined to inform S&T investments toward delivering and measuring capabilities and their effectiveness. However, without metrics and methods to characterize S&T investments, planners have no way of measuring operational effects or determining how best to impose complexity and thus shape adversary behavior.

Our research involved a literature review of complexity, graph theory, and CAS; historical case studies of warfare and competition; internal workshops with wargamers, complexity, and C2 subject-matter experts and sponsors; and leveraging emerging MDO CONOPs from recent wargames. Based on this research, we developed four example vignettes.

We applied the findings of our study of CAS to four warfare and operational vignettes to understand how to leverage complexity to U.S. advantage. Based on our research, we binned CAS characteristics into two categories. The first represents CAS's characteristics associated with the adversary C2 system and processes and provides a definitional basis for identifying complex C2 systems that might be influenced by complexity attacks. The second set of characteristics identifies the mechanisms that can destabilize complex systems. These mechanisms describe the avenues that enable CAS-informed actions or attacks on the adversary's decision processes.

Building on this analysis, we presented notional analyses that demonstrate the application of complexity attacks to operational vignettes based on the ACE concept and the SEAD mission. To provide a grounded modeling and analytical strategy, we show that U.S. actions on the adversary's decision points in a military vignette can be represented by CAS characteristics depicted in a Markov chain (see Volume II for details). Understanding the underlying transition matrix structure (or associated directed graph) provides a framework for analyzing the operational impacts produced by complexity exploitation. More challenging is developing a model with defined probabilities in the Markov chain, which would be estimated through intelligence, wargaming, or simulation but that would provide additional insights for warfighting by offering a grounded framework for comparing the operational impacts of different multidomain efforts.

To take advantage of opportunities to leverage complexity in operational settings, the Air Force should apply a complexity lens to review existing and planned efforts in terms of complexity. The lens should be applied for offensive opportunities against the adversary's decision processes, and it should be used to review U.S. systems to understand vulnerabilities and mitigation measures. Figure 2.1 provided CAS mapping to warfighting that military planners can use to identify properties of CAS that may be associated with the adversary's C2 (and, hence, vulnerabilities) and avenues for complexity exploitation that can be used to assess alternative courses of action under consideration. The planning of MDOs and evaluation of MDO effectiveness are areas where grounded analysis based on the complexity lens might help shape choices to apply multidomain effects for military advantage. The lens would also help identify linkages between efforts that may help advocate for continued funding and pursuit to truly leverage capabilities across domains. The S&T research agenda of the military labs can also be evaluated through the complexity lens; the framing of a complexity lens application to S&T is discussed in the companion volume.

To advance research in complexity in warfare, AFRL should conduct game theory–informed wargames or subject-matter expert workshops and surveys to populate probabilities in the Markov chain representation of the adversary's complex decision process. To guide the employment of these concepts in an operational setting , the Air Force should conduct workshops that leverage the complexity lens to inform joint all-domain operations. Similarly, Pacific Air Forces and U.S. Air Forces in Europe should fold the complexity lens thinking into existing tabletop and command post exercises to weigh the value of military deception and complexity imposition strategies as part of deterrence or kinetic strategies.

References

Air Force, *Air Force Future Operating Concept*, Washington, D.C., September 2015. As of
March 18, 2020:
https://www.af.mil/Portals/1/images/airpower/AFFOC.pdf

Brinsmead, Thomas S., and Cliff Hooker, "Complex Systems Dynamics and Sustainability:
Conception, Method and Policy," in Cliff Hooker, ed., *Philosophy of Complex Systems*,
Vol. 10 (*Handbook of the Philosophy of Science*), Amsterdam, North Holland: Elsevier,
2011, pp. 809–838.

Churchman, Charles West, "Guest Editorial: Wicked Problems," *Management Science*, Vol. 14,
No. 4, December 1967, pp. B141–B142.

Diehl, Ernst, and John D. Sterman, "Effects of Feedback Complexity on Dynamic Decision
Making," *Organizational Behavior and Human Decision Processes*, Vol. 62, No. 2, 1995,
pp. 198–125.

Engstrom, Jeffery, *Systems Confrontation and System Destruction Warfare: How the Chinese
People's Liberation Army Seeks to Wage Modern Warfare*, Santa Monica, Calif.: RAND
Corporation, RR-1708-OSD, 2018. As of January 8, 2021:
https://www.rand.org/pubs/research_reports/RR1708.html

Estrada, Ernesto, *The Structure of Complex Networks: Theory and Applications*, New York:
Oxford University Press, 2011.

Garfinkel, Alan, Jane Shevtsov, and Yina Guo, *Modeling Life: The Mathematics of Biological
Systems*, Cham, Switzerland: Springer International Publishing AG, 2017.

Grösser, Stefan N., "Complexity Management and System Dynamics Thinking," in Stefan N.
Grösser, Arcadio Reyes-Lecuona, and Göran Granholm, eds., *Dynamics of Long-Life
Assets*, Cham, Switzerland: Springer, 2017, pp. 69–92.

Hofstetter, Dominic, "Innovating in Complexity (Part II): From Single-Point Solutions to
Directional Systems Innovation," *Medium*, July 25, 2019. As of January 18, 2021:
https://medium.com/in-search-of-leverage/innovating-in-complexity-part-ii-from-single-
point-solutions-to-directional-systems-innovation-dfb36fcfe50

Kwakkel, Jan H., Marjolijn Haasnoot, and Warren Walker, "Developing Dynamic Adaptive
Policy Pathways: A Computer-Assisted Approach for Developing Adaptive Strategies for a
Deeply Uncertain World," *Climatic Change*, Vol. 132, 2015, pp. 373–386.

Lempert, Robert J., "A New Decision Sciences for Complex Systems," *Proceedings of the National Academy of Sciences*, Vol. 99, Supp. 3, May 2002, pp. 7309–7313.

Lempert, Robert J., Steven W. Popper, and Steven C. Bankes, *Shaping the Next One Hundred Years: New Methods for Quantitative, Long-Term Policy Analysis*, Santa Monica, Calif.: RAND Corporation, MR-1626-RPC, 2003. As of January 18, 2021:
https://www.rand.org/pubs/monograph_reports/MR1626.html

Lempert, Robert J., Steven W. Popper, David G. Groves, Nidhi Kalra, Jordan R. Fischbach, Steven C. Bankes, Benjamin P. Bryant, Myles T. Collins, Klaus Keller, Andrew Hackbarth, et al., *Making Good Decisions Without Predictions: Robust Decision Making for Planning Under Deep Uncertainty*, Santa Monica, Calif.: RAND Corporation, RB-9701, 2013. As of January 18, 2021:
https://www.rand.org/pubs/research_briefs/RB9701.html

Lingel, Sherrill, Matthew Sargent, Tim Gulden, Tim McDonald, and Parousia Rockstroh, *Leveraging Complexity in Great-Power Competition and Warfare*: Volume II, *Technical Details for a Complex Adaptive Systems Lens*, Santa Monica, Calif.: RAND Corporation, RR-A589-2, 2021. As of August 2021:
https://www.rand.org/t/RRA589-2.html

Lynn, William J., III, "Defending a New Domain: The Pentagon's Cyberstrategy," *Foreign Affairs*, Vol. 89, No. 5, September/October 2010, pp. 97–108.

Marchau, Vincent, Warren Walker, Pieter Bloemen, and Steven Popper, eds., *Decisionmaking Under Deep Uncertainty: From Theory to Practice*, Cham, Switzerland: Springer, 2019.

Mazzocchi, Fulvio, "Complexity in Biology: Exceeding the Limits of Reductionism and Determinism Using Complexity Theory," *European Molecular Biology Organization*, Vol. 9, No. 1, January 2008, pp. 10–14.

Meadows, Donella, *Leverage Points: Places to Intervene in a System*, Hartland, Vt.: The Sustainability Institute, 1999. As of January 18, 2021:
http://donellameadows.org/wp-content/userfiles/Leverage_Points.pdf

Mitchell, Melanie, *Complexity: A Guided Tour*, Oxford: Oxford University Press, 2009.

Mitchell, Sandra D., *Unsimple Truths: Science, Complexity, and Policy*, Chicago: University of Chicago Press, 2009.

Morris, Lyle J., Michael J. Mazarr, Jeffrey W. Hornung, Stephanie Pezard, Anika Binnendijk, and Marta Kepe, *Gaining Competitive Advantage in the Gray Zone: Response Options for Coercive Aggression Below the Threshold of Major War*, Santa Monica, Calif.: RAND Corporation, RR-2942-OSD, 2019. As of January 18, 2021:
https://www.rand.org/pubs/research_reports/RR2942.html

Murphy, Eric M., *Complex Adaptive Systems and the Development of Force Structures for the United States Air Force*, Drew Paper No. 18, Maxwell Air Force Base, Ala.: Air University Press, Air Force Research Institute, 2014. As of January 8, 2021: https://www.airuniversity.af.edu/Portals/10/AUPress/Papers/DP_0018_MURPHY_COMPLEX_ADAPTIVE_SYSTEMS.PDF

National Academies of Sciences, Engineering, and Medicine, *Guiding Cancer Control: A Path to Transformation*, Washington, D.C.: National Academies Press, 2019.

Richardson, Michael J., Alexandra Paxton, and Nikita Kuznetsov, "Nonlinear Methods for Understanding Complex Dynamical Phenomena in Psychological Science," *Psychological Science Agenda*, February 2017. As of January 8, 2021: https://www.apa.org/science/about/psa/2017/02/dynamical-phenomena

Rickles, Dean, Penelope Hawe, and Alan Shiell, "A Simple Guide to Chaos and Complexity," *Journal of Epidemiology Community Health*, Vol. 61, No. 11, November 2007, pp. 933–937.

Rittel, Horst W. J., and Melvin M. Webber, "Dilemmas in a General Theory of Planning," *Policy Sciences*, Vol. 4, 1973, pp. 155–169.

Rosenthal, Uriel, R. Arjen Boin, and Louise K. Comfort, "The Changing World of Crises and Crises Management," in Uriel Rosenthal, Arjen Boin, and Louise K. Comfort, eds., *Managing Crises*, Springfield, Ill.: Charles C. Thomas Publishers Inc., 2001, pp. 5–27.

Ruiz-Martin, Cristina, Adolfo López Paredes, and Gabriel A. Wainer, "Applying Complex Network Theory to the Assessment of Organizational Resilience," *IFAC-PapersOnLine*, Vol. 48, No. 3, 2015, pp. 1224–1229.

Russell, Martha G., and Nataliya V. Smorodinskaya, "Leveraging Complexity for Ecosystemic Innovation," *Technological Forecasting and Social Change*, Vol. 136, November 2018, pp. 114–131.

Simon, Herbert A., "Theories of Bounded Rationality," in C. B. McGuire and Roy Radner, eds., *Decision and Organization*, Amsterdam: North-Holland Publishing Company, 1972, pp. 161–176.

U.S. Department of Defense, "Remarks by Secretary Mattis on the National Defense Strategy," webpage, January 19, 2018. As of March 17, 2020: https://www.defense.gov/Newsroom/Transcripts/Transcript/Article/1420042/remarks-by-secretary-mattis-on-the-national-defense-strategy/

Watts, Barry D., *The Military Use of Space: A Diagnostic Assessment*, Washington, D.C.: Center for Strategic and Budgetary Assessments, February 2001, pp. 5–14.